WATCH ME PRACTICE

Math Workbook

Grade 2

Watch Me Practice Grade 2 Math Workbook

For more information about this title or to order other books and/or electronic media, contact the publisher:

B.L. Academic Services, LLC

http://watchmepracticeseries.com

info@watchmepracticeseries.com

First Edition

ISBN: 978-1-7369237-1-9

Printed in the United States

Cover Designer: Jamira Ink Designs LLC

Interior Designer/Editor: 1106 Design

Table of Contents

COUNTING TO 100

In Grade 1, you learned to count to 50. Now, count from 50 to 100 by writing the numbers 50 through 100.
For example:

50	51	52	53	54	55	56	57	58	59
60									

Write the numbers 50 through 100 in words. **For example:**

Fifty	Fifty-one	Fifty-two	Fifty-three	Fifty-four	Fifty-five	Fifty-six	Fifty-seven	Fifty-eight	Fifty-nine
Sixty									

SKIP COUNTING TO 100

1. Skip count by twos to 20

   2, _____, _____, _____, 10, _____, _____, _____, _____, 20.

2. Skip count by threes to 33

   3, _____, _____, _____, _____, _____, _____, _____, _____, _____, 33

3. Skip count by fours to 40

   4, _____, _____, _____, _____, _____, _____, _____, _____, 40

4. Skip count by fives to 50

   5, _____, _____, _____, _____, _____, _____, _____, _____, 50

5. Skip count by sixes to 66

   6, _____, _____, _____, _____, _____, _____, _____, _____, _____, 66

6. Skip count by sevens to 77

   7, _____, _____, _____, _____, _____, _____, _____, _____, _____, 77

7. Skip count by eights to 88

   8, _____, _____, _____, _____, _____, _____, _____, _____, _____, 88

8. Skip count by nines to 99

   9, _____, _____, _____, _____, _____, _____, _____, _____, _____, 99

9. Skip count by tens to 100

   10, _____, _____, _____, _____, _____, _____, _____, _____, 100

COUNTING TO 1000

When counting to 1,000, you can skip count by tens or hundreds.

10	20	30	40	50	60	70	80	90	100
110	120	130	140	150	160	170	180	190	200
210									

Directions:
First, complete the chart by skip counting by tens to get to 1,000.
Next, point to the column that skip counts by 100 to get to 1,000.

EVEN NUMBERS AND ODD NUMBERS

Even numbers end with the digits 0, 2, 4, 6, or 8. Odd numbers end with the digits 1, 3, 5, 7, or 9.
For example:

    a. 38 is an even number, because it ends with the digit 8

    b. 37 is an odd number, because it ends with the digit 7

Even Numbers or Odd Numbers

Circle the even numbers. Draw a heart around the odd numbers.

        17,   34,   92,   105,   76,   55,   63,   40,   111,   222,   81,   29,

Write a three-digit even number. _____

Write a three-digit odd number. _____

Difference Between Even Numbers and Odd Numbers

Even numbers can be divided into two groups evenly. Odd numbers cannot be divided into two groups evenly. **For example:**

    a. A group of eight marbles can be evenly divided into two groups:

$$(O \quad O \quad O \quad O) \quad (O \quad O \quad O \quad O)$$

    b. A group of nine marbles cannot be evenly divided into two groups:

$$(O \quad O \quad O \quad O) \quad (O \quad O \quad O \quad O) \quad O$$

    1. Divide the 16 marbles into two groups:

        O O O O O O O O O O O O O O O O

    2. How would you divide nineteen marbles between two people?

        O O O O O O O O O O O O O O O O O O O

    3. Name the next even number that comes after 42. _____

    4. Name the last odd number that comes before 53. _____

    5. Name the even number that comes between 72 and 76. _____

    6. Name the odd number that comes between 85 and 89. _____

PLACE VALUE

The digits in a number have a place value. If a number has one digit, that number is in the ones place. If a number has two digits, the digit on the right is in the ones place, and the digit on the left is in the tens place. **For example:**

The number 25 has two digits. The 2 is in the tens place, and the 5 is in the ones place. Two tens = 20.   Five ones = 5.   20 + 5 = 25.

Complete.

    1. The number 38 has _____ tens and _____ ones.

If a number has three digits, the digit on the left is in the hundreds place, the digit in the middle is in the tens place, and the digit on the right is in the ones place.

For example: The number 164 has three digits. The 1 is in the hundreds place, the 6 is in the tens place, and the 4 is in the ones place.

Complete.

    2. The number 279 has _____ hundreds, _____ tens, and _____ ones.

Place Value and Expanded Form

Digits in a number have a place value. For example, 45 has 4 tens and 5 ones.

$$4 \text{ tens} = 10 + 10 + 10 + 10 = 40$$
$$5 \text{ ones} = 1 + 1 + 1 + 1 + 1 = 5$$
$$40 + 5 = 45$$

When you write the number 45 as 40 + 5, it is called **expanded form.** Expanded form is writing a number as the place value of its digits.

For example: 164 = 100 + 60 + 4

Write the following numbers in expanded form.

    25 _____

    173 _____

    509 _____

GREATER THAN, LESS THAN, EQUAL TO

The greater than sign looks like this >
The less than sign looks like this <
The equal sign looks like this =
The greater than sign opens to the larger number. **For example: 8 > 6.**
The less than sign points to the lesser number. **For example: 6 < 8.**
The equal sign means both numbers are the same. **For example: 9 = 9.**

Use the >, <, or = sign to complete the following:

1. 3 is _____ 2

2. 5 is _____ 10

3. 12 is _____ 12

4. 4 + 7 is _____ 12 + 4

5. 13 – 7 is _____ 14 – 7

6. 20 + 1 is _____ 20 + 5

7. 31 – 1 is _____ 30 – 5

8. 500 is _____ 400

9. 673 is _____ 763

10. 420 – 20 is _____ 400

ROUNDING

When you estimate, you don't give an exact answer. You say about how many.
Rounding helps you to estimate.

Rules for Rounding to the Nearest Ten:

When rounding a number to the nearest ten, follow these steps:

First, look at the digit in the ones place. If the digit in the ones place is between 1 and 4, round down to zero.

Next, if the digit in the ones place is between 5 and 9, round the number in the tens place up to the next digit, and change the digit in the ones place to zero.

For example: Round 32 to the nearest ten.

What is the digit in the ones place? _____
Is the digit in the ones place between 1 and 4 or 5 and 9? _____
What should you do?

Hint: If the digit in the ones place is between 1 and 4, round down to zero. So, the 2 becomes a 0.

Answer: 30

Estimate the sum of 25 + 14 =
  a. First round 25 to the nearest ten by looking at the digit in the ones place. The digit in the ones place is 5. So, round the digit in the tens place up to the next digit, which is 3, and change the digit in the ones place to zero. 25 rounded to the nearest ten becomes 30.

  b. Next, round 14 to the nearest ten by looking at the digit in the ones place. The digit in the ones place is 4. So, round the digit in the ones place down to zero. The number 14 rounded to the nearest ten becomes 10.

  c. Last, the estimated sum of 25 + 14 becomes 30 + 10 = 40.

Round to the nearest ten
  1. Round 15 to the nearest ten. _____
  2. Round 22 to the nearest ten. _____
  3. Round 39 to the nearest ten. _____

Estimate the following sums and differences by rounding to the nearest ten.

1.	48 + 51	**2.**	63 + 27	**3.**	79 + 38
4.	84 − 42	**5.**	95 − 57	**6.**	28 − 12

ORDINAL NUMBERS

Ordinal numbers tell the position of numbers. Ordinal numbers also tell the position of grades in school.

For example:

Grade 1 = First grade or 1st grade
Grade 2 = Second grade or 2nd grade

Write the remaining position of grades from Grade 3 through Grade 12

Grade	Ordinal Number	Abbreviation
Grade 1	First Grade	1st Grade
Grade 2	Second Grade	2nd Grade
Grade 3		
Grade 4		
Grade 5		
Grade 6		
Grade 7		
Grade 8		
Grade 9		
Grade 10		
Grade 11		
Grade 12		

Ordinal numbers also tell the position of days on a calendar.
For example, on a calendar during the month of June:

a. Day 13 can be written as the thirteenth day of June, or abbreviated, as the 13th day of June.

b. Day 14 can be written as the fourteenth day of June, or abbreviated, as the 14th day of June.

Using the following table, write the days of the month through the thirty-first using ordinal numbers.

1st	2nd			

ADDITION

Addition vocabulary words:

1. **Sum** — the answer you get when adding numbers together

2. **Addends** — the numbers that are added together to get a sum

For example:

$10 + 8 = 18$

The **addends** are 10 and 8.
18 is the **sum**.

Complete.

1. $9 + 4 + 1 =$ _____

2. $7 + 5 =$ _____

3. How many addends are in problem **1**? _____

4. How many addends are in problem **2**? _____

5. Write an addition problem that has three addends that equal the sum of 11. _____

Adding Numbers in Any Order

No matter the order of the addends, the sum will always be the same.
For example: $5 + 4 = 9$ AND $4 + 5 = 9$

Add

1. $10 + 3 =$ _____

2. $3 + 10 =$ _____

3. $15 + 5 =$ _____

4. $5 + 15 =$ _____

Checking Your Addition Answers

You can check your addition answers by subtracting.

For example: 8 + 8 = 16

Check your answer by using the sum to subtract one of the addends: 16 − 8 = 8

Find the sums. Then check your answers using subtraction.

| **1.** | 33
+ 4 | **2.** | 42
+ 7 | **3.** | 15
+ 4 | **4.** | 25
+ 2 |

Check your answers.

1.   **2.**   **3.**   **4.**

Addition with Two-Digit Numbers

When adding numbers with two digits, add the digits in the ones place first.
Then add the digits in the tens place.

For example: 24
                 +13   Check your answer using subtraction: 37
                  37   −13
                                                        24

When adding 24 and 13, add the digits in the ones column first, in this case 4 + 3.
Next, add the digits in the tens column, in this case 2 + 1. The sum is 37.

Add

	1.	2.	3.
	67	92	43
	+ 22	+34	+ 35

Check your answers using subtraction:

1.　　　　　　　2.　　　　　　　3.

Adding Two-Digit Numbers with Regrouping

When adding two-digit numbers, you may have to regroup.

For example:　47　　Check your answer using subtraction:　65
　　　　　　　+ 18　　　　　　　　　　　　　　　　　　　　−18
　　　　　　　───　　　　　　　　　　　　　　　　　　　　───
　　　　　　　65　　　　　　　　　　　　　　　　　　　　　47

When adding 47 and 18, add the digits in the ones column first, which, in this case, is 7 + 8. When you add 7 and 8, you get a sum of 15. You must regroup 15 by placing the 1 above the 4 in the tens column and placing the 5 below the 8 in the ones column. There are now three digits in the tens column. Add all three digits to get a sum of 6. The sum of 47 + 18 = 65.

To check your answer, subtract the sum by using one of the addends. In the example, the addend 18 was used. To subtract the sum of 65 using the addend 18, you will have to regroup when checking your answer because 8 cannot be subtracted from 5. Regroup by borrowing from the 6 in the tens column. The 5 ones become 15. The 6 tens become 5 tens. The difference between 65 − 18 = 47.

Add.

1. 55
   + 27

2. 82
   + 39

3. 74
   + 46

Check your answers using subtraction.

1.

2.

3.

Adding Three Two-Digit Numbers with and without Regrouping

When adding three two-digit numbers, add the digits in the ones column first. Next, add the digits in the tens column. **For example:**

without regrouping	with regrouping
21	33
30	44
+48	+ 55
99	132

a. First add the digits in the ones column. Regroup if needed.

b. Next, add the digits in the tens column.

Add

1. 35
   14
   + 26

2. 42
   30
   + 58

3. 65
   22
   + 74

4. 74
   46
   + 27

Addition with Three-Digit Numbers with and without Regrouping

When adding three-digit numbers with and without regrouping, follow the same order as when adding two-digit numbers.

For example:

```
  354          Check your answer using subtraction:   480
+ 126                                                -126
-----                                                -----
  480                                                  354
```

a. First, add the digits in the ones column. Regroup if needed.

b. Next, add the digits in the tens column. Regroup if needed.

c. Then, add the digits in the hundreds column.

Add

1.
```
  432
+ 211
-----
```

2.
```
  652
+ 309
-----
```

3.
```
  274
+ 921
-----
```

4.
```
  142
+ 744
-----
```

Check your answers using subtraction.

1.   2.   3.   4.

SUBTRACTION

Subtraction vocabulary words:

1. Difference — the answer you get when you subtract numbers.

2. Subtrahend — the numbers that are being subtracted.

For example: 8 − 2 = 6. The difference is 6. The subtrahends are 8 and 2.

Checking Your Subtraction Answers

You can check your subtraction answers by adding.
For example: 8 - 2 = 6

Check your answer by adding the difference to the lesser subtrahend.

For example: 6 (difference) +2 (lesser subtrahend) = 8

Find the differences:

1. What is the difference of 10 minus 7? _____

2. What is the difference of 12 minus 6? _____

3. How many subtrahends are in the following subtraction problem?

   8 − 6 − 2 = 0 _____

4. Write a subtraction problem that has two subtrahends with a difference of 4.

Subtraction with Two-Digits

When subtracting numbers with two digits, subtract the digits in the ones column first. Next, subtract the digits in the tens column. **For example:**

$$\begin{array}{r} 24 \\ -13 \\ \hline 11 \end{array}$$

1. Subtract the digits in the ones column, 4 minus 3.

2. Next, subtract the digits in the tens column, 2 minus 1.

3. The difference of 24 minus 13 is 11.

Checking Your Subtraction Answers with Two-Digits

You can check your subtraction answers with two-digits by adding.

For example: 24 – 13 = 11

Check your answer by adding the difference to the lesser subtrahend.

For example: 11 (difference) +13 (lesser subtrahend) = 24

Subtract

1.	2.	3.	4.
67	92	85	65
−50	−10	−34	−22

When subtracting numbers with two digits, you may have to regroup. **For example:**

$$\begin{array}{r} 47 \\ -18 \\ \hline 29 \end{array}$$

a. When subtracting numbers with two digits, subtract the digits in the ones column first. In this case, the 8 cannot be subtracted from 7, so you must regroup. Borrow from the 4 in the tens column, leaving 3 tens. Regroup by adding the borrowed number with the 7 in the ones column. The 7 becomes 17. You are now able to subtract 8 from 17.

b. Next, subtract the digits in the tens column.

c. The difference of 47 minus 18 = 29.

Subtract

1.	72	2.	58	3.	34	4.	25
	−44		−39		−27		−16

Checking Your Subtraction Answers with and without Regrouping

When checking your answer to a two-digit subtraction problem, change the subtraction problem to an addition problem. **For example:**

Without regrouping	**Check your answer using addition:**
78	54
−24	+24
54	78

Check your answer by adding the difference to the lesser subtrahend.

With regrouping

$$\begin{array}{r} 53 \\ -16 \\ \hline 37 \end{array}$$

Check your answer using addition:

$$\begin{array}{r} 37 \\ +16 \\ \hline 53 \end{array}$$

Check your answer by adding the difference to the lesser subtrahend.

Subtract. Then, check your answers using addition.

1. $\begin{array}{r} 88 \\ -74 \\ \hline \end{array}$ 2. $\begin{array}{r} 92 \\ -17 \\ \hline \end{array}$ 3. $\begin{array}{r} 47 \\ -18 \\ \hline \end{array}$ 4. $\begin{array}{r} 53 \\ -16 \\ \hline \end{array}$

Check your answers using addition.

1.   2.   3.   4.

Subtraction with Three-Digit Numbers with and without Regrouping

When subtracting numbers with three digits, subtract the digits in the ones column first. Regroup if needed. Next, subtract the digits in the tens column. Regroup if needed. Last, subtract the digits in the hundreds column.

For example:
$$\begin{array}{r} 523 \\ -247 \\ \hline 276 \end{array}$$

Check your answer using addition:
$$\begin{array}{r} 276 \\ +247 \\ \hline 523 \end{array}$$

Check your answer by adding the difference to the lesser subtrahend.

Subtract. Then, check your answers using addition.

1. $\begin{array}{r} 622 \\ -536 \\ \hline \end{array}$

2. $\begin{array}{r} 239 \\ -156 \\ \hline \end{array}$

3. $\begin{array}{r} 487 \\ -226 \\ \hline \end{array}$

4. $\begin{array}{r} 509 \\ -468 \\ \hline \end{array}$

Check your answers using addition.

1.

2.

3.

4.

Coins

Say the names of the following coins. Next, in each column, tell what each coin is worth using words, a cent sign, and a dollar sign. **For example:**

Penny	**one cent**	**1¢**	**$0.01**
Nickel			
Dime			
Quarter			

There is a coin that is worth fifty cents. It is called a half dollar because fifty cents is half of $1.00. A half dollar can be written as $0.50, fifty cents, or 50¢.

Complete.

1. How many pennies would you need to make $0.25? _____ .

2. How many pennies would you need to make $1.00? _____ .

3. Name three coins that would equal $0.75. _____

4. How many quarters make up $1.00? _____

5. How many nickels would you need to make 35¢? _____ .

Adding Money

When adding money, follow the same order as when adding two-digit and three-digit numbers with and without regrouping. However, when adding money, you must include the decimal point behind the dollar amount in your answer. **For example:**

Without regrouping

$1.25
+ $0.10
——
$1.35

Check your answer using subtraction:

$1.35
– $0.10
——
$1.25

With regrouping

$2.25
+ $3.15
——
$5.40

Check your answer using subtraction:

$5.40
– $3.15
——
$2.25

Add. Then, check your answers using subtraction.

1. $10.58
   +$12.21

2. $23.14
   +$19.50

3. $37.90
   +$46.77

Check your answers using subtraction.

1.

2.

3.

Subtracting Money

When subtracting money, follow the same order as when subtracting two-digit and three-digit numbers with and without regrouping. When subtracting money, you must also include the decimal point behind the dollar amount in your answer. **For example:**

Without regrouping

$$\begin{array}{r} \$5.23 \\ -\$2.11 \\ \hline \$3.12 \end{array}$$

Check your answer using addition:

$$\begin{array}{r} \$3.12 \\ +\$2.11 \\ \hline \$5.23 \end{array}$$

With regrouping

$$\begin{array}{r} \$3.15 \\ -\$2.25 \\ \hline \$0.90 \end{array}$$

Check your answer using addition:

$$\begin{array}{r} \$0.90 \\ +\$2.25 \\ \hline \$3.15 \end{array}$$

Subtract. Then, check your answers using addition.

1.
$$\begin{array}{r} \$12.21 \\ -\$10.58 \\ \hline \end{array}$$

2.
$$\begin{array}{r} \$23.14 \\ -\$19.50 \\ \hline \end{array}$$

3.
$$\begin{array}{r} \$46.77 \\ -\$37.90 \\ \hline \end{array}$$

Check your answers using addition.

1.

2.

3.

MULTIPLICATION

Multiplication vocabulary words.

1. Factor — the numbers you are multiplying.

2. Product — the answer you get after multiplying factors.

Multiplication is a quick way of adding numbers by skip counting. Like adding, multiplication increases a number. Skip counting can help you memorize multiplication facts. For example, when memorizing the multiplication facts for 2, skip count by two.

2 x 0 =	0
2 x 1 =	2
2 x 2 =	4
2 x 3 =	6
2 x 4 =	8
2 x 5 =	10
2 x 6 =	12
2 x 7 =	14
2 x 8 =	16
2 x 9 =	18
2 x 10 =	20
2 x 11 =	22

Things to remember when multiplying:

a. A factor multiplied by zero is always zero.

b. A factor multiplied by one is always that factor.

Practice finding the product of the following factors:

Factor 0	Factor 1	Factor 3	Factor 4	Factor 5
0 x 0 =	1 x 0 =	3 x 0 =	4 x 0 =	5 x 0 =
0 x 1 =	1 x 1 =	3 x 1 =	4 x 1 =	5 x 1 =
0 x 2 =	1 x 2 =	3 x 2 =	4 x 2 =	5 x 2 =
0 x 3 =	1 x 3 =	3 x 3 =	4 x 3 =	5 x 3 =
0 x 4 =	1 x 4 =	3 x 4 =	4 x 4 =	5 x 4 =
0 x 5 =	1 x 5 =	3 x 5 =	4 x 5 =	5 x 5 =
0 x 6 =	1 x 6 =	3 x 6 =	4 x 6 =	5 x 6 =
0 x 7 =	1 x 7 =	3 x 7 =	4 x 7 =	5 x 7 =
0 x 8 =	1 x 8 =	3 x 8 =	4 x 8 =	5 x 8 =
0 x 9 =	1 x 9 =	3 x 9 =	4 x 9 =	5 x 9 =
0 x 10 =	1 x 10 =	3 x 10 =	4 x 10 =	5 x 10 =

Multiplication Word Problems

1. What two factors would Angelica have to multiply to get the product of 4? _____

2. Juar had 3 baseball cards. Quinton had twice as many baseball cards. How many baseball cards did Quinton have? _____

3. What two factors would Kayla have to multiply to get the product of 45? _____

Factor 6	Factor 7	Factor 8	Factor 9	Factor 10
6 x 0 =	7 x 0 =	8 x 0 =	9 x 0 =	10 x 0 =
6 x 1 =	7 x 1 =	8 x 1 =	9 x 1 =	10 x 1 =
6 x 2 =	7 x 2 =	8 x 2 =	9 x 2 =	10 x 2 =
6 x 3 =	7 x 3 =	8 x 3 =	9 x 3 =	10 x 3 =
6 x 4 =	7 x 4 =	8 x 4 =	9 x 4 =	10 x 4 =
6 x 5 =	7 x 5 =	8 x 5 =	9 x 5 =	10 x 5 =
6 x 6 =	7 x 6 =	8 x 6 =	9 x 6 =	10 x 6 =
6 x 7 =	7 x 7 =	8 x 7 =	9 x 7 =	10 x 7 =
6 x 8 =	7 x 8 =	8 x 8 =	9 x 8 =	10 x 8 =
6 x 9 =	7 x 9 =	8 x 9 =	9 x 9 =	10 x 9 =
6 x 10 =	7 x 10 =	8 x 10 =	9 x 10 =	10 x 10 =

Multiplication Word Problem

4. Kendall wants to pass out lollipops to her classmates for her birthday. There are 25 students in her class. There are five lollipops in each bag. How many bags of lollipops would Kendall have to buy so that each of her classmates gets a lollipop? _____

DIVISION

Division vocabulary words.

1. Dividend — number that is being divided.

2. Divisor — number that does the dividing.

3. Quotient — the answer you get after dividing the dividend.

Division is a quick way of breaking down numbers into groups. When you think of dividing, think of dividing objects into groups. A division problem has three parts: the dividend, the divisor, and the quotient.

For example:

Eight divided by two can be written as $8 \div 2 = 4$.
The 8 is the dividend because it is the number being divided.
The 2 is the divisor because it is the number doing the dividing.
The 4 is the quotient because $8 \div 2 = 4$.

Division Rules:

a. Dividing zero by a number is always zero. For example: $0 \div 7 = 0$

b. Dividing a number by 1 always equals that number. For example: $3 \div 1 = 3$

Practice memorizing the division facts 2 through 5.

Divisor of 2	Divisor of 3	Divisor of 4	Divisor of 5
$0 \div 2 = 0$	$0 \div 3 = 0$	$0 \div 4 = 0$	$0 \div 5 = 0$
$2 \div 2 = 1$	$3 \div 3 = 1$	$4 \div 4 = 1$	$5 \div 5 = 1$
$4 \div 2 = 2$	$6 \div 3 = 2$	$8 \div 4 = 2$	$10 \div 5 = 2$
$6 \div 2 = 3$	$9 \div 3 = 3$	$12 \div 4 = 3$	$15 \div 5 = 3$
$8 \div 2 = 4$	$12 \div 3 = 4$	$16 \div 4 = 4$	$20 \div 5 = 4$
$10 \div 2 = 5$	$15 \div 3 = 5$	$20 \div 4 = 5$	$25 \div 5 = 5$
$12 \div 2 = 6$	$18 \div 3 = 6$	$24 \div 4 = 6$	$30 \div 5 = 6$
$14 \div 2 = 7$	$21 \div 3 = 7$	$28 \div 4 = 7$	$35 \div 5 = 7$
$16 \div 2 = 8$	$24 \div 3 = 8$	$32 \div 4 = 8$	$40 \div 5 = 8$
$18 \div 2 = 9$	$27 \div 3 = 9$	$36 \div 4 = 9$	$45 \div 5 = 9$

There are two kinds of clocks. The digital clock shows the time using numbers. The analog clock tells time using the hour hand and the minute hand. On an analog clock, the short hand is called the hour hand. The long hand is called the minute hand. The minute hand tells how much time passes between the numbers on the clock skip counting by fives and ones. **For example:**

   a.   When the minute hand passes from the 12 to the 1, five minutes have passed.

   b.   When the minute hand passes from the 12 to the 2, ten minutes have passed.

Look at an analog clock at home to help you answer the following questions.

   1.   If the hour hand is on the 9 and the minute hand is on the 3, what time would it be?   _____

   2.   If the hour hand is on the 12 and the minute hand is on the 5, what time would it be?   _____

Telling Time to the Half-Hour

The minute hand tells how much time passes between the numbers on a clock skip counting by fives. When the minute hand is on the number 6, we say it is 30 minutes past or half past the hour. **For example:**

If the hour hand is on the 10 and the minute hand is on the 6, we can say what time it is in three ways: 10:30, thirty minutes past ten, or half past 10.

Look at an analog clock at home to help you answer the following question.

   a.   If the hour hand is on the 2 and the minute hand is on the 6, name three ways you could say what time it would be.

   _____

   _____

   _____

Telling Time using A.M. or P.M.

When telling time, you must know if it is in the morning, afternoon, or evening. There are a total of 24 hours in a day.

1. A.M. tells us the time in the morning. There are 12 hours in the a.m. 12 a.m. – 12 p.m.

2. P.M. tells us the time in the afternoon or evening. There are 12 hours in the p.m. 12 p.m. – 12 a.m.

For example:

a. If it is 10 o'clock in the morning, we can write the time as 10 a.m.

b. If it is 10 o'clock in the evening, we can write the time as 10 p.m.

Answer the following questions using a.m. or p.m.

1. It is 2:30 in the afternoon. How would you write the time?

   _____

2. It is 3 o'clock in the morning. How would you write the time?

   _____

USING A CALENDAR

On a calendar, time is given by days of the month, the year, days of the week, and the date.

List the months that have 31 days	List the months that have 30 days	List the month that has fewer than 30 days

USING A BAR GRAPH

This bar graph shows favorite ice-cream flavors of 2nd-grade students at The Brown Elementary School. Bar graphs are used to compare things.

Use the bar graph below to answer the questions.

Ice Cream Flavors

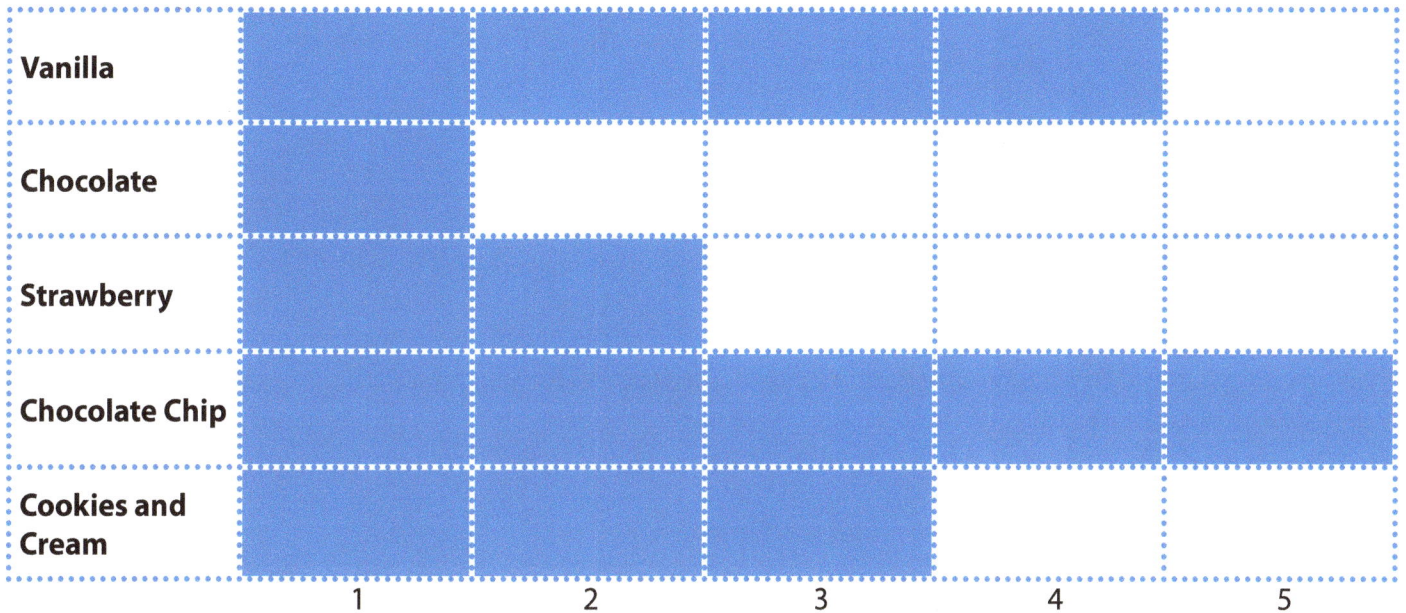

	1	2	3	4	5
Vanilla					
Chocolate					
Strawberry					
Chocolate Chip					
Cookies and Cream					

1. Which ice cream is the least favorite? _____

2. Which ice cream is the favorite? _____

3. How many students like strawberry ice cream? _____

4. How many students like cookies and cream? _____

5. Which two flavors combined gives a total of 9? _____

6. What does this graph tell us about the ice cream that 2nd-graders like? _____

FRACTIONS

A fraction is made up of a numerator and a denominator. The numerator tells how much of the whole part is being used. The denominator gives the whole part.

For example:

1   The top number is the numerator
—
2   The bottom number is the denominator

There are two parts of this figure. One part is shaded. The fraction for this figure is one-half or ½ because one part of the figure is being used.

Answer the following questions below about the figure.

1. How many parts are in the figure? _____ . How many parts are shaded? _____

2. What would the numerator be? _____

3. What would the denominator be? _____

4. How would you write the fraction? _____